AROUND THE BAY

CENTER FOR LAND USE INTERPRETATION · AMERICAN REGIONAL LANDSCAPE SERIES

AROUND THE BAY

Man-Made Sites of Interest in the San Francisco Bay Region

Blast Books
NEW YORK

PREFACE

San Francisco emerged, famously, with the Gold Rush and over the following century became America's portal to the Pacific world. In this period the Bay's shoreline developed as a regionwide port for defense and offense, with the most transformative period during World War II. In the postwar years the region sprawled, like Los Angeles. Today the nine-county Bay Area has more than 7 million residents. It's a megacity, with a big wet hole in the middle.

The Bay itself can be viewed as a geographic paradox: a place and a void. The collective "Bay" (composed of San Francisco Bay, San Pablo Bay, and Suisun Bay) both unites and divides the community of the Bay Area, giving identity to a region while separating its populace. The Bay is a back space where the hardened surfaces of the industrial city crumble into the water, as well as a shore front with designed parks and recreational marinas. It is intensely visited in places and nearly inaccessible in others; its beauty acclaimed, its dumping grounds unparalleled; its sparkling water refreshed from Sierra snowmelt, its sewer outfalls and urban runoff robust. Once intensely militarized, it is now, just as intensely, *de*militarized. In a sense, the Bay is a natural entity, borne of great rivers draining the entire Central Valley of California; however, every inch of its shoreline today is the product of human activity, by either intent or incident.

PRESIDIO

As the principal mouth allowing access into the body of California, the Golden Gate was heavily fortified on both sides, with batteries and bases on the Headlands and at Fort Baker to the north and at the Presidio on the south. The Presidio was a military city within the city, covering hundreds of acres and housing thousands. It was established by the Spanish in 1776, the year of U.S. independence, and was occupied by U.S. forces in 1846, in time for the Gold Rush. When it closed in 1994, it had been the longest continuously active military base in the nation. Most of its land now belongs to the National Park Service and the Presidio Trust, mandated to achieve financial self-sufficiency for the property by 2013. This was accomplished by redevelopment and leasing many of its eight hundred buildings to a variety of public and private entities, including the Internet Archive, the Walt Disney Family Museum, and the film-effects company Industrial Light and Magic.

FISHERMAN'S WHARF/AQUATIC PARK

San Francisco's maritime history is preserved and interpreted at Aquatic Park Cove, a shoreline area protected by the great arch of the Municipal Pier. Features of this part of the shoreline include the art deco ship-shaped Maritime Museum building, a collection of historic vessels docked at the Hyde Street Pier, and a World War II ship and submarine parked at Pier 45. Also located here are the ferries, restaurants, and docks of Fisherman's Wharf, as this neighborhood is called, and where a small fleet of active fishing boats is still based. At the far end is Pier 39, a popular shopping and dining attraction that opened in 1978, which helped turn this part of the shore into one of the most touristed places in the nation.

BAY BRIDGE WEST LANDING

The San Francisco–Oakland Bay Bridge connects on the San Francisco side at the South of Market neighborhood, after spanning Pier 26 and the Embarcadero, and touches ground next to a new condominium tower. This marks the western end of Interstate 80, after a 2,899-mile journey westward from the George Washington Bridge in New York City. The elevated highway divides the city, establishing a transition from the high-rises of downtown and the financial district to the relatively flat expanses of the former industrial and logistics area to the south. The shoreline around downtown has been significantly extended with fill over the years. During the excavation of the site for the Infinity Condominiums here in 2005, for example, three blocks from the current shoreline, the remains of the *Candace*, an early China trade and whaling ship built in Boston in 1818, was unearthed.

MISSION BAY

South of downtown San Francisco and the AT&T Park along Mission Creek, the paved expanses of the former railyards at Mission Bay are evolving into a large master-planned redevelopment project. The remnants of a shipping and industrial center for San Francisco (industries that formed the city in the nineteenth and twentieth centuries) are being converted into corporate office parks and residential areas, with a University of San Francisco research campus in the middle—a brownfield redevelopment reflecting the industries of the twenty-first century. Protruding from the shore is the Mission Rock Terminal, which, like much of the region's shoreline, is part of the Port of San Francisco. U.S. Maritime Administration transport roll-on/roll-off ships are usually docked here, ready to be deployed within a few days' notice in the event of an emergency. Next to the terminal is an abandoned railroad ferry slip that used to bring freight cars from across the Bay, and across the country, from a similar slip at Point Richmond.

POTRERO POINT

The largest intact industrial port site in San Francisco, Potrero Point is one of the oldest continuously operating civilian shipyards in the country. Union Iron Works moved to this 23-acre location in 1881 and employed some thirteen hundred people to build ships, bridges, mining machinery, and weapons for use in developing the West. The yard was later bought by Bethlehem Steel, which built dozens of destroyers and repaired submarines here during World War II. BAE Systems, a British military contractor, now operates the main ship repair facilities, known as Pier 70, including the largest floating dry dock on the West Coast, where cruise ships can often be seen undergoing repairs. Potrero Point is also the location of the police department's auto impoundment yard and the last power plant to operate within the city.

INDIA BASIN

Once a container yard, the expanses of Piers 96 and 94 are now used occasionally for an emergency vehicle operations training course and for parking a reserve cargo ship (the Port of Oakland, across the Bay, long ago captured most of the container traffic in the region). A large shed overhangs the shore, built for a novel cargo handling method known as a LASH system, which used floating containers that could be loaded with goods and floated onto and out of oceangoing ships, easing trade at shallow ports and into the narrow canals of the world. This method, favored for a while in Europe, has mostly been discontinued. Across from the shed is Heron's Head Park, a promontory of earth that was to be the site for a bridge across San Francisco Bay that was never built. Also known as Pier 98, this is the southern limit of Port of San Francisco property.

HUNTERS POINT

Hunters Point, a heavily industrialized and contaminated U.S. Navy shipyard, is in the midst of the slow conversion to civilian use. It operated as a military shipyard from 1941 to 1974, servicing ships of all kinds, from submarines to aircraft carriers. After Pacific Island nuclear bomb tests in the 1940s, irradiated ships were brought back to Hunters Point for decontamination, and the Naval Radiological Defense Lab was established to develop defenses against the effects of exposure to radiation sustained during a nuclear attack. The legacy of this organization, and the mountains of radioactive material produced and handled here, is one of the lingering concerns about the base. Soil remediation and removal at the site is extensive, and some burial pits have been known to smolder.

CANDLESTICK POINT

Candlestick Point is a former shoreline dumping ground bought by California in 1973 and turned into the first state-owned urban recreation area. It is dominated by Candlestick Park, a 65,000-seat stadium, surrounded by the largest parking lot in San Francisco, with 10,000 spaces. The stadium was built in 1960 as part of a deal to bring the Giants baseball team from New York. The team played there for forty years, moving to the new PacBell/AT&T arena in 2000. The local NFL football team, the 49ers, has used the arena as its home field since 1970. In the future the team will move into a new stadium being built for it in Santa Clara, at the southern end of the Bay. Candlestick Park will likely be torn down within months after the end of the 2013 football season and the site redeveloped with thousands of new homes.

BRISBANE LANDFILL

This part of Brisbane was a major railroad storage and repair site, the Bayshore Railyards, until the 1980s, and the primary dumping ground for San Francisco's trash for several decades. Dumping started with the abundant supply of fill material created by the 1906 earthquake, when the depositing of debris from the quake transformed the entire shoreline around the Bay, making hundreds of acres of new land. Dumping continued into this former marsh in Brisbane into the 1960s, when the landfill was officially closed. San Francisco's main waste transfer station is still located at the north end of the former dump site. This is also the location of the nation's first official artist-in-residence program based at a municipal waste site. Artists are granted picking rights to the trash as it comes in. Although not technically a landfill anymore, loads of excavated material continue to be added to the low mounds next to the highway, to accelerate the compaction of the decomposing trash underneath it.

SAN FRANCISCO AIRPORT

Airports are often among the first major construction projects on marshes and estuaries, and San Francisco International Airport is no exception, built on hundreds of acres of former cow pasture and landfill that extends into the Bay. SFO is a major gateway to the Pacific and the second busiest airport in California, after LAX. Reflecting the decline in the dot-com economy, the airport's use dropped by 10 million passengers in less than three years, bottoming out at fewer than 30 million passengers in 2003. Its use has picked up since then, and it is now back to handling more than 40 million passengers a year, ranking it around the twenty-fifth busiest airport in the world. The proximity of the parallel runways to one another limits its capacity. Recent efforts to expand the runway areas into the Bay have been thwarted by environmental concerns.

FOSTER CITY

Generally considered to be the northern limit of Silicon Valley, Foster City is an old masterplanned community built on former marshland in the mid-1960s. It is named after the real estate developer who conceived it, Jack Foster. The city is 20 square miles in size, although 16 square miles of it is water—engineered ponds and channels as well as land submerged by the Bay. It is primarily residential, with a current population of thirty thousand and a median household income around $135,000. Biotech companies located here employ more than three thousand, and Visa, the credit card company founded in Fresno and headquartered on Market Street in San Francisco, employs more than a thousand in its main offices at Foster City. The San Mateo Bridge, crossing the Bay to Hayward, lands in Foster City and, at 7 miles in length, is the longest bridge in the Bay Area; it was said to be the longest bridge in the world when it opened in 1929.

REDWOOD SHORES

Like Foster City, its more notorious neighbor to the north, 1,500-acre master-planned Redwood Shores was conceived by a single company in the 1960s and built in an undeveloped area that was more bay than land. Also like Foster City, a network of artificial lagoons was built into the plan, so that many of the houses have frontage on the water. Redwood Shores was started by the owners of the site, the Leslie Salt Company, which operated an evaporation pond on the premises. Leslie sold the stalled development to the Mobil Oil Company in 1973. It is now largely built out, with home prices exceeding $1 million, and the community is now part of Redwood City. At its southwestern end is the headquarters for the Oracle Corporation, built in 1989, after Marine World moved from the site to Vallejo. At the northeastern end, on the Bay, is a sewage treatment plant and two powerful broadcasting facilities, one that once beamed Christian programming around the world via shortwave and another, known as the "50,000-watt flamethrower," that broadcasts sports programming on KNBR AM 680.

PORT OF REDWOOD CITY

The Port of Redwood City is the only deepwater port in the South Bay. It developed as early as the 1850s as a loading area farther up Redwood Creek for the redwood trees that were harvested from the hills and taken up to San Francisco as construction material. Slowly land was filled in and the port area extended farther out into the marshlands of the Bay, and the Army Corps of Engineers kept the waterway dredged for ships. Bulk industries continue to dominate the port, mostly material recycling, including construction debris, recycled industrial fuel, concrete, asphalt, and scrap metal. Gypsum comes from Mexico to a terminal here for wallboard. Cargill Salt also uses the area for salt crystallization and harvesting, though most of this activity takes place across the Bay in Newark. Affluent high-tech development recently encroached in the form of the new Pacific Shores Center, an office complex that surrounds a sports center at the end of the peninsula, with tenants that include DreamWorks Animation studios and a new yacht harbor.

DUMBARTON BRIDGE/HETCH HETCHY AQUEDUCT

A number of important and historic water crossings are clustered here in the South Bay. The Dumbarton Bridge was the first road bridge to span the Bay, and today it connects the built-out Silicon Valley to the real estate of the southeast Bay. The train bridge south of the Dumbarton Bridge was the first rail bridge across the Bay, the Dumbarton cutoff, opening in 1910. Service stopped in 1982, and the rotating span, which allows boats through, is now welded open. A submerged pipeline next to the bridge moved salt brine to the ponds and the refinery at Redwood City. The Hetch Hetchy Aqueduct, a major source of water for San Francisco that brings water from the famous reservoir in the Sierras, makes its Bay crossing here in two large silver pipes. Tall power lines along the bridge bring Hetch Hetchy electricity from the Newark substation to the Ravenswood substation and then north to the city. Next to the Ravenswood substation is a structural collapse training site operated by the fire department. Beyond it is the main campus for Facebook. Unusually shaped islands were constructed in a former salt evaporation pond to restore bird habitat.

COOLEY LANDING

This built-up protrusion into the Bay at East Palo Alto started in 1849 as Ravenswood Wharf, for a while the only port between San Francisco and San Jose. It developed into a port for shipping bricks throughout the area, produced at a plant onshore, a site that evolved into an industrial area that continues to house scrap yards, hazardous waste, and substations. The site was bought in 1868 by Lester Cooley, and it became known as Cooley Landing. From 1932 to 1960 it was a county dump, after which it became the Palo Alto Boat Works, a boatyard specializing in the repair of small wooden boats. The owner of the yard, Carl Schoof, sold it to the Peninsula Open Space Trust in 1999 for wildlife preservation, and it is being converted into a park. One building from the old boatyard has been preserved.

BYXBEE PARK

Palo Alto meets the Bay in an interesting collection of terminal sites. An active landfill for the city lies next to the wastewater treatment plant for the region, which discharges into the adjacent slough. When the runway for the local airport was built, the small yacht harbor's drainage was hampered, and the harbor silted up and was abandoned, although some of the old club buildings remain next to an aerated duck pond. Land art constructed on the closed portions of the landfill, called Byxbee Park, include a collection of sculpted mounds, poles, K-rail, and other structures. Behind the landfill/park is an abandoned radio station, a relay site for KFS, one of the last commercial shortwave broadcasting facilities to operate in the nation, which transmitted its final radiotelegraph message in July 1999 in Morse code, using the words Samuel Morse uttered 155 years prior, at the invention of the telegraph: "What hath God wrought."

GOOGLEPLEX

This former shoreline landfill in Mountain View that was turned into a park and surrounded by a golf course (built on top of a landfill) is itself surrounded by the offices of Google, known as Googleplex, and includes the original central campus, which Google bought from Silicon Graphics. The tentlike stage of the Shoreline Amphitheatre is also located here, a performance space that opened in 1986. It was designed by the legendary concert promoter Bill Graham and evokes the shape of the Grateful Dead's skull graphic from the cover of the band's live album *Steal Your Face*. Graham died five years later when his helicopter crashed into electrical wires in San Pablo Bay on his way back from a Huey Lewis concert. During the stage's first year of operation, concertgoers were known to use cigarette lighters to ignite methane gas leaking from the landfill underneath them.

MOFFETT FIELD

Many of the high-tech industries of the South Bay developed around the military's demands for surveillance. Moffett Field was a military base established in the 1930s, when blimps served an important role as surveillance platforms. Later technologies, such as radar, high-altitude aircraft, and satellites, made them obsolete. Before the navy left Moffett Field in 1994, it was the nation's primary base for P-3 submarine-spotting airplanes. In the 1950s NACA (later NASA) developed a major R & D center at the base, called Ames Research Center, which focused on aeronautic engineering and is associated with Dryden Research Center at Edwards Air Force Base. Ames still operates the base and maintains the R & D facilities, including a large assortment of wind tunnels. Hangar One, the largest of three old blimp hangars on base, covers 8 acres. The airfield still has military tenants, including an army psychological operations group.

LOCKHEED MARTIN

Lockheed Martin Space Systems in Sunnyvale, adjacent to Moffett Field, is one of the most important satellite development and manufacturing plants in the United States. Established in 1957 as Lockheed's Missile Systems Center, then developing the Polaris submarine-launched ICBM, the facility covers 412 acres and employs a few thousand people. In addition to developing several types of rockets, more than 850 orbited satellites have been built here, including civilian imaging satellites such as Landsat 7 and many communications and broadcast satellites. Lockheed is the nation's largest defense contractor, and many of the satellites created here are projects for the defense community, such as Milstar and the Corona program. At the southwest corner of the site is Onizuka Air Force Station, known generally as the Blue Cube, once one of the military's busiest surveillance satellite earth stations. It was used by the National Reconnaissance Office starting in 1961, until those functions were moved to Colorado in 2007. In 2011 most of the Cube's remaining functions were transferred to Vandenberg Air Force Base. At the northern end of the Lockheed plant is Yahoo!'s corporate campus.

ALVISO

Located at the bottom of the San Francisco Bay, Alviso is an old port town with a dried-up marina. More than a hundred years ago, Alviso was San Jose's shipping port, and after railways made the port less vital it became a major cannery town. As a result of the heavy draw on the groundwater of San Jose, Alviso has sunk an estimated 15 feet and is now protected by levees and pumps. Although Silicon Valley's office park developments are encroaching, the town retains some of its original character, and some buildings more than 150 years old still stand despite a number of historic floods. Among them, the Bayside Cannery building was part of one of the largest canneries in the nation in the 1920s. The area is within the EPA's South Bay Asbestos Area Superfund site, part of the legacy of landfills in the region. Some levees made with asbestos-contaminated soil have recently been replaced.

SAN JOSE TREATMENT PLANT

Every community generates sewage that needs to be treated, and San Francisco Bay is ringed with dozens of sewage treatment plants, all of which discharge into the Bay. The largest of these is the San Jose Water Pollution Control Plant, which treats the effluent from most of Santa Clara County, serving a population of more than 1.5 million people. Although it operated for many years as only a primary treatment plant, the facility now has secondary and tertiary systems, which make the discharge reasonably clean where it flows through artesian slough and into the southern end of the Bay, though there has been some discussion about extending the outfall to the Dumbarton Bridge. The sludge drying ponds next to the plant cover nearly 2 square miles.

NEWBY ISLAND LANDFILL

The Newby Landfill, started in 1938, is the largest active dump on the shores of the Bay. It is the terminus for waste for San Jose, Milpitas, and other cities. The 342-acre pile is still shy of its permitted height of 120 feet and has years to go before it is scheduled to close. The landfill is an island surrounded by a levee that keeps its runoff from directly entering the Bay, and the water that drains from it is treated in the dump's own treatment plant. Electricity for the dump is generated by burning the methane collected from the decomposition of the waste. In 2012 one of the largest trash-sorting facilities in the nation opened here, where dozens of people sort through waste coming in on conveyors, removing material that can be recycled or reused. What remains goes into the landfill. Dried sewage sludge from the nearby San Jose treatment plant is used as cover, mixed in with the trash, blending San Jose's waste streams.

MUD SLOUGH

Mud Slough is a channel that runs through the colorful salt evaporation ponds (still owned by Cargill and used to make salt at its production center in Newark) and drains into the southern portion of Fremont, now the fourth largest city in the Bay Area. Fremont is where some technology companies from Silicon Valley, across the Bay, have expanded. The Tesla Motors facility is a large industrial plant formerly owned by GM and Toyota, where the company is producing the Model S electric car, its highly anticipated fully electric sedan with a range greater than 250 miles. Delivery started in 2012, though numbers fell far short of the five thousand cars the company said it would produce in 2012. Just south of Tesla is another notorious green energy site, the production center for the Solyndra solar panel company, notorious for going bankrupt in 2011 after receiving a $535 million dollar loan from the U.S. Department of Energy.

TRI-CITIES LANDFILL

The Tri-Cities Landfill closed in 2011 after more than forty years of accepting waste from Union City, Fremont, and Newark. Waste for the region now goes to a transfer facility nearby where recyclables are extracted, and the remaining trash is hauled to Waste Management Incorporated's large regional megafill near Altamont, in the hills east of Livermore, 30 miles away. That landfill accepts around 1 million tons of trash per year and is permitted past the year 2030. The Bay Area's trash-shed expands outward, merging with its energy-shed, which also depends on the region's hinterlands for production. Behind the Tri-Cities Landfill mound is the Newark substation, owned by PG&E, which operates all electrical distribution in the region. This substation steps voltage down from high-voltage transmission current out of the region's major power sources, which include several gas-fired plants on Suisun Bay, to a lower, regional transmission voltage for distribution to San Francisco, Oakland, and San Jose.

CARGILL SALT

Once covering more than 40,000 acres, the salt operations at the southern end of the San Francisco Bay are among the largest in the country, though they have been recently reduced to about 12,000 acres. Having bought the Bay Area's main salt producer, Leslie Salt, in 1978, Cargill (a privately held industrial conglomerate with more than a hundred thousand employees worldwide) owns all salt operations on the Bay. The salt production system is in two parts, with some harvesting done in Redwood City and the majority taking place around this plant in Newark. Salt brine is shifted from pond to pond over as long as a five-year period, until it is saline enough to grow as a crystal layer, about 12 inches thick, on the bottom of the crystallizing ponds. It is then scooped up from the bottom of these ponds and is washed, ground, and packaged. Most of the salt is used for roads, chemicals, and the food industry (including specialty pellets made here for the Campbell's Soup Company). Only 3 percent ends up as table salt. The ponds are red in color due to naturally growing halophilic bacteria.

COYOTE HILLS

The Coyote Hills are unusual because they are a natural formation, unlike most shoreline hills of the South Bay that were artificially formed by piling refuse. The Dumbarton Quarry at the southern end of the hill, next to the toll plaza for the Dumbarton Bridge, is as deep as the hills are high, which makes for an unusual shift in elevation, from 300 feet above sea level, to sea level, to 300 feet below sea level. The DeSilva Company, owners of the quarry, agreed to cease operations here in 2007 in exchange for an expansion of its operations in the hills above Fremont, and the process of converting the hole into a freshwater lake was begun. The lake is part of a regional park, which includes a former Nike missile site—one of a dozen in the Bay Area built in the 1960s and soon abandoned (and one of four Nike sites now in the East Bay Regional Park District). Also on site is the Patterson reservoir, a covered tank that holds 14 million gallons of drinking water.

EDEN LANDING

The Union Sanitary District plant is the primary wastewater treatment facility for the "tri-cities" region of Union City, Newark, and Fremont, serving the needs of about 328,000 people. It is located in the sectional salt pond area near the mouth of Alameda Creek, the drainage channel for the largest watershed in the South Bay. The creek flows through a series of flooded quarries in Fremont, known as the Quarry Lakes, then into the hills through Niles Canyon and Sunol, draining the mountains east of San Jose. At the Bay, the creek discharges into an area known as the Eden Landing Ecological Reserve, some 6,000 acres of former salt ponds south of the San Mateo Bridge now managed by the California Department of Fish and Wildlife. A model-aircraft flying field is located next to the wastewater plant, operated by the Southern Alameda County Radio Controllers Club. Miniature airfields for remote-control aircraft clubs are often found in flat marginal areas, near sewer plants and flood-control basins.

OYSTER BAY

Oyster Bay Regional Shoreline is a waste confluence in San Leandro, at the southern end of Oakland Airport. The circular mound landform is a former landfill in the Bay, used for about forty years, then closed in 1979. It is still undergoing treatment and grading to manage runoff and reduce contamination from hazardous materials, although it is open to the public and contains some outdoor sculpture. Waste Management Incorporated operates a large waste-sorting and transfer facility at the site, with as many as seven hundred trucks arriving every day delivering 2 million pounds of trash to sort and load back on trucks headed to landfills at Altamont and beyond. The city of San Leandro has its wastewater treatment plant here too, which discharges secondary-treated sewage into the Bay and dries sludge in piles behind the plant. Next to that is a golf course recently reinstalled after being used as a dump site for dredge spoils from a deepening project at the Port of Oakland. An indoor shooting range is here as well, another common use of land on urban margins, operated by the local chapter of the Optimist Club.

OAKLAND AIRPORT

The second busiest of the three large international airports in the Bay Area, Oakland serves about 10 million passengers a year, a quarter of the traffic of SFO. Some 70 percent of its traffic comes from Southwest Airlines, which sends about a hundred flights a day from the airport. It is also a major FedEx hub. The airport opened in 1927, with the longest runway in the world at the time. Amelia Earhart took off from here in 1937 on her final flight, when she disappeared somewhere over the Pacific. In 1962 the airport was expanded with a new 10,000-foot runway and taxiway built entirely on bay fill, west of the existing airport. The old part of the airport is still used for civil aviation but is a bit of a relic.

COAST GUARD ISLAND

Coast Guard Island, known officially as of 2012 as Base Alameda, is a 67-acre island in the Oakland Estuary. The island was formed from dredging projects in 1913 and developed into a training and supply center, becoming the largest Coast Guard field unit on the West Coast in the 1960s. It was known as Government Island until it was renamed in 1982, when the training-center function moved to New Jersey. Coast Guard Island serves chiefly as an administrative center for personnel and base-support functions throughout the Pacific, including procurement, industrial support, and housing. It is the home of the Coast Guard's Pacific Area command and home port for oceangoing ships, including the largest in the Coast Guard's fleet, the 418-foot-long National Security Cutters used for patrolling the world's oceans. The Coast Guard is now one of the five armed services of the United States (along with the marines, army, navy, and air force), but it is the only official military force in the Department of Homeland Security.

ALAMEDA NAVAL AIR STATION SITE

The U.S. Navy began building the 1,734-acre base on Alameda Island in the late 1930s, and for more than fifty years it was a repair and maintenance facility for Navy aircraft, including carrier-based planes and helicopters. It was closed in 1997 and is now in the lengthy transition stage from a military base to a civilian extension of the city of Alameda. Several military ships still dock at the station, mostly part of the MARAD reserve fleet, and one aircraft carrier is now a museum. Large naval seaplanes were stationed here, and their ramps and hangars remain. As with other recently closed military sites in the Bay Area, among the first major users of the hangars and closed runways is the film industry, which often occupies these between places with abundant space and few restrictions on use. The locally produced *MythBusters* TV show regularly uses the asphalt expanses for field tests. Remains of a mile-long highway set built for one of the *Matrix* movies remains visible. Otherwise more than a full square mile of abandoned runways and munitions bunkers, with views across the Bay to San Francisco, are fenced off from public use, enjoyed by nesting birds, while lingering ground contamination is slowly being addressed.

PORT OF OAKLAND

Oakland is the fifth most active port in the United States and the second largest container port on the West Coast (after L.A./Long Beach), due in large part to Oakland being the western terminus for major transcontinental railways. A substantial expansion of the port took place recently using land freed up from the closure of military supply bases in the port area—much of the port was a former depot and shipping area for the army and navy from World War II to the early 1990s. The dock terminal areas were recently dredged to a depth of 50 feet to accommodate a new generation of larger container ships, and more asphalt yards and container cranes were added (there are now more than thirty). Oakland was the first West Coast port for Sea-Land's containerized system, developed in the 1960s, which became the standard method for international "box" shipping.

BAY BRIDGE EAST LANDING

The Oakland side of the Bay Bridge is a dynamic landscape of transit for waste, radio, cargo, rail, and automobiles. The Bay Bridge, though upstaged by its flashy neighbor at the Golden Gate (which opened in 1937, six months after the Bay Bridge), has more superlatives associated with it: for many years the Bay Bridge was the longest bridge in the world as well as the largest steel structure in the world; its central anchorage was the most massive object in the world and its toll plaza the widest in the world. It is the busiest toll bridge in the nation, averaging 270,000 cars per day and generating $8 million a month. The eastern half, from Yerba Buena Island to Oakland, is being replaced after its partial collapse in the 1989 earthquake. The eastern span replacement will cost around $6.5 billion and be completed in 2013, after more than ten years' construction. Both east and west spans of the original bridge took three and a half years to build and cost $77 million—about $1.3 billion in 2013 dollars.

BERKELEY PIER

A number of distinct protrusions extend from the shoreline along the East Bay, such as the marina at Emeryville and this one at Berkeley. What is known as the Berkeley Pier is now a 3,000-foot fishing pier extending from the shore at the Berkeley Marina, but this pier predates the marina. It was built in 1926 for an auto ferry, taking cars across the Bay to San Francisco, serving as the last leg of what used to be U.S. Highway 40, which ended at University Avenue, on the water's edge. The auto pier was 3.5 miles long, crossing halfway across the Bay, nearly to where Treasure Island is today. It was made obsolete when the Bay Bridge opened in 1936, and it was left to rot. Much of it is still there, a distinct hazard to navigation. The bulk of the shoreline protrusion at Berkeley was built starting in the 1930s, absorbing the base of the pier, made of debris and general refuse. The marina portion was built at that time as a WPA project, along with the nearby Aquatic Park. The trapezoidal north appendage was made through municipal waste landfilling from the 1950s into the 1980s, after which it was capped with soil and turned into North Waterfront Park, later renamed Cesar Chavez Park.

ALBANY DUMP AND POINT ISABEL

These two peninsulas are former landfills. Point Isabel has some big box–type architecture (a Costco and a major junk-mail processing facility) as well as a storm-runoff water treatment plant, a dog park, and a radio broadcast station. The undeveloped Albany Dump peninsula (known as the Bulb) is composed of twisted metal, broken concrete, and other fragments of the urban landscape dumped into the Bay up until 1987. It has become a creative sculpture park and homeless campsite. At the base of the peninsula is Golden Gate Fields, the Bay Area's largest horse racing track, bought from the Catellus Corporation by an English gaming company. The track is built on the site of a former explosives plant that blew up for the last time in 1892, shattering windows in San Francisco.

RICHMOND FIELD STATION

The University of California has operated an engineering field station on a portion of this former chemical industry site on the Bay since 1950, conducting building engineering testing, forestry products research, transportation studies, and environmental research at a number of unique facilities on the site. These include an earthquake simulation building and a former fog chamber and flight simulation film production building, now used for pavement research. The university uses some of the original buildings from the California Blasting Cap Company, which was the former owner of the property. Historically the site has been used as a copper refinery, pesticide plant, and for the production of sulfuric acid and other things at the former AstraZeneca plant next door, most of which has recently been torn down. The extent of the pollution on site is still being assessed.

PORT OF RICHMOND

Richmond's port is the third largest port in California (after L.A./Long Beach and Oakland) when measured in tonnage of commodities handled annually. Like the town itself, the port is a product of World War II and oil. Although the principal channel, the Santa Fe, began to be dredged in the 1920s, it wasn't until Henry Kaiser built four shipyards at the port during the war that the infrastructure was laid for the port as we see it today. Richmond was the largest wartime shipbuilding operation on the West Coast, producing more than seven hundred ships. Bulk commodities such gypsum, petroleum, and scrap metal later grew to dominate the port area. Now new car distribution is the primary activity on the port's west side, where more than two hundred thousand cars are off-loaded from Japan and Korea annually. This is one of Honda's principal ports of entry to the United States. Protecting the port from waves is a long breakwater and Brooks Island, once a quarry and a private hunt club used by the likes of Bing Crosby.

POINT RICHMOND

Point Richmond is an area that is accessible primarily through a tunnel from the inland community of Point Richmond, and it is a sort of microcosm of past and future forms of development along the Bay. As the western terminus of the Santa Fe Railway, in 1900 this area was a busy shipping terminal on the Bay. A shoreline warehouse, built in 1915 as part of the early port, is still there. Next to it, the ferry terminal too remains, where rail cars were shipped across the Bay to China Basin in San Francisco. A major brick plant, one of several on the Bay that were busy during the rebuilding of the region following the 1906 earthquake, has now been turned into a condominium complex, called Brickyard Cove, and new homes have been built on manufactured peninsulas on lots that were technically sold as submerged land.

CASTRO POINT

Ferries were once a major part of life in the Bay Area, a form of public transportation that brought much of the population into direct contact with the Bay on a daily basis. Castro Point, at the base of the Richmond–San Rafael Bridge, vividly shows the transition between the Bay as a public medium to the Bay as a thing to be surmounted. Ferries ran continuously from this point to San Quentin Point until 1956, when the "largest continuous steel bridge in the world" was built, 5.5 miles long, connecting Marin County to Contra Costa County and the East Bay. The bridge recently underwent a $500 million seismic retrofit. The remains of the ferry terminal still crumble next to the bridge, complete with a collection of sunken hulks lurking offshore. Ships were sunk here to create a breakwater and harbor for the Red Rock Marina, which no longer exists. Red Rock Island, a rocky 5-acre mound just offshore, has been listed for sale for years, with no takers yet. West of the bridge is Long Wharf, the port for the Chevron Richmond Refinery, where a 36-inch-diameter pipe moves up to 10 million gallons of crude per day off tankers from Alaska and the Middle East.

POINT MOLATE

Point Molate was developed by the California Wine Association in 1908 as a central winery for processing grapes from all over the state. The Winehaven Winery, as it was called, became the largest winery in the United States, producing 12 million gallons of wine and port per year at its peak before being shut down during Prohibition (though it continued to make sacramental wine until 1937). In 1941 the U.S. Navy purchased the 400-acre property for use as a fuel supply depot, turning the castlelike winery building into a headquarters and occupying the historic housing. Most of the fuel was kept in twenty underground concrete tanks, with a capacity of more than 40 million gallons, built on the hillside above the winery. The facility was officially closed in 1998, and the City of Richmond is in the process of taking it over, once cleanup issues are resolved, with plans to develop the site.

POINT SAN PABLO

In 1950 the shores of Point San Pablo were lined with docks and reduction plants for animal processing, such as the production of fish oils and tallow. The last rendering plant burned down around 1990, by which time the facilities at the site were used largely for storage and logistics for chemical industries. Metal tanks on the hill stored everything from ammonia to sulfuric acid to molasses. The site, known as Terminal 4 of the Port of Richmond, is being cleaned up by the last tenant, the Paktank company, part of the large Dutch chemical distribution conglomerate Vopak. Ruins of what may have been the last commercial whaling station in the United States, shut down by court order in 1970, remain along the shore past the tip of the point. Farther down is the Point San Pablo Yacht Harbor, a small recreational port protected by a harbor made by sinking six old schooners and a ferryboat. Just outside the port is the submerged remains of a steamer sunk during the filming of *Blood Alley*, a 1955 John Wayne movie.

RICHMOND REFINERY

The Chevron Richmond Refinery is one of the largest and oldest refineries on the West Coast. Its construction started in 1901, and it was soon bought by Standard Oil. The refinery covers 2,900 acres, has 5,000 miles of pipelines, and features dozens of large tanks that can hold up to 15 million barrels of crude, gasoline, jet fuel, diesel, lube oil, wax, and other chemicals produced by the refinery. Of the sixty thousand who work for Chevron worldwide, sixteen hundred people are employed here. A second California refinery operated by the company is in the Los Angeles community of El Segundo. The Chevron name was chosen when the company bought Gulf Oil in 1984, in what was at the time one of the largest corporate mergers in history. The company doubled in size after buying Texaco in 2001 and Unocal in 2005. Chevron was based in San Francisco for nearly a century but moved in 2001 to a new corporate campus in the quiet Bay Area suburb of San Ramon.

WEST CONTRA COSTA COUNTY LANDFILL

The West Contra Costa County Landfill in North Richmond is one of the largest and oldest continuously active landfills on the Bay. It ceased accepting municipal waste in 2006, although it still serves as a compost and waste processing site and as a transfer station for shipping regional trash farther inland (currently to the Keller Canyon Landfill near Pittsburg). It serves communities all along the south shore of San Pablo Bay, from Crockett to parts of Berkeley. The pile that started in 1952 as bay fill has reached its permitted terminal height of 235 feet. Because of compaction caused by the weight of the mound, the waste extends 25 feet below sea level. Like most landfills in the Bay Area, WCCC is privately owned, most recently by Republic Services, the second largest waste hauler in the nation (after WMI), which in addition to owning around two hundred landfills across the country also owns the National and Alamo car rental chains and the used car chain AutoNation. The landfill still accepts dried sludge from the county sewage treatment plant located next door. The adjacent water treatment facility pipes its liquid effluent to the Chevron refinery, which evaporates it in cooling plants.

POINT PINOLE

Point Pinole is a 2,100-acre park on the site of a large explosives plant. In 1892, after disastrous explosions at two San Francisco sites and another at the relocated plant in Albany (the current site of Golden Gate Fields), the Giant Powder Company relocated here and stayed until 1960. Giant was the first American company licensed to use Alfred Nobel's newly patented product: dynamite. In 1915 Giant was bought by the Atlas Powder Company. With the invention of ammonium nitrate explosives in the 1940s, the plant slowly became obsolete. When the plant closed in 1961, NASA considered building its Mission Control Center here but decided instead on Houston. Walt Disney bought the narrow-gauge railway that once moved explosives around and installed it at his new amusement park, Disneyland. Bethlehem Steel acquired the land in the early 1960s and operated a structural steel galvanizing plant here for more than a decade. The land was sold and became a park opened to the public in 1973, with explosion deflection berms and some other structures remaining from the powder plant.

HERCULES

In 1893 California Powder Works moved its black powder production from Santa Cruz to this then remote area, up the shore from the new Giant Powder Company plant at Pinole. The company, later renamed Hercules, was one of the first companies to be formed by judicial decree, as part of the breakup of DuPont's monopoly in the explosives industry. By 1917 it was the most productive TNT plant in the United States. With the invention of nitroglycerine, Hercules evolved into a chemical and fertilizer company, producing formaldehyde, methanol, and anhydrous ammonia. Other plants owned by the company went on to produce rocket motors for ICBMs. The plant closed in 1977 and the land around the plant began to be developed into a residential area and an office park. The old headquarters building, with foot-thick walls to protect executives in case of an accidental explosion, is still standing, as are some other buildings, the remains of the old wharf, and piles of contaminated dirt.

RODEO REFINERY

The refinery in Rodeo was the first of the five major oil refineries now operating on the shores of the Bay Area. Built in 1896, the plant now covers more than 1,000 acres and employs five hundred people. It processes 100,000 barrels of crude per day, chiefly to make gasoline. For years it was owned by Unocal but was bought by Tosco in 1997, and then by ConocoPhillips, which spun off its Phillips 66 division as a separate company in 2012. The re-formed Phillips 66 company operates a dozen refineries in the United States and has 10,0000 branded gas stations and 15,000 miles of pipelines. The Rodeo refinery is connected to a smaller refinery in Arroyo Grande, California, near Pismo Beach, 200 miles south, via a pipeline.

SELBY WORKS SITE

A large slab of asphalt on the shore in Rodeo is the protective cap covering the contaminated remains of the Selby Smelting Works. After operating here for more than eighty-five years, the gold, silver, and lead smelter was closed and demolished, along with its company town Tormey, in 1971. The Selby Works moved here in 1884 from San Francisco, where its belching stacks were visible and well known, to be farther from the city, closer to a deepwater port, and able to expand its operations. It became the largest smelter on the West Coast and was operated for its last sixty-five years by the giant American Smelting and Refining Company, ASARCO. The 610-foot-tall brick chimney was once the tallest in the world. The expense of upgrading the plant to conform to new environmental regulations, such as the Clean Air Act of 1970, and an expanding suburban population led to the company's decision to close the plant.

CARQUINEZ BRIDGE

The Bay system is at its narrowest point at the western end of Carquinez Strait, a 6-mile-long submerged canyon that separates San Pablo Bay from Suisun Bay. The Bay Area's bridges have many superlatives associated with them, and the Carquinez Bridge is no exception. It was called the highest bridge in the world when it was built and for many years was called the largest cantilever bridge in the United States. The bridge now consists of two parallel bridges, one made in 1958 for the interstate and another built a few years ago. The new bridge, used by westbound I-80, replaced the original 1927 bridge located in the strait between the two newer bridges, torn down when the new one opened in 2003. The town of Crockett grew as a wheat port in the late 1800s, and what is now the C&H (California and Hawaii) sugar plant was originally a flour mill. Today, after more than a hundred years, the sugar plant is still likely one of the largest sugarcane refineries in the world, processing a few million pounds of cane daily, brought by ship from the company's plantations in Hawaii.

CARQUINEZ STRAIT

Because of its narrowness and the volume of water flushing through it, the Carquinez Strait's waters are some of the deepest in the Bay Area. With the completion of the rail line from Martinez, this became an important deepwater port for the inland region, including Central Valley farms, especially productive in wheat in the 1880s. At that time, half the ships heading out the Golden Gate were loaded with wheat from these docks, bound for Europe. The remains of nearly 4 miles of grain wharves lining the strait are still visible. Later, the railway was continued through to the west, and a bridge was built from Benicia, ending the boom for this area. The once bustling waterfront "wheat port" town of Port Costa is a relic from this period and has found favor as a motorcycling destination.

MARTINEZ

Martinez is an industrial town with two major oil refineries, a few chemical plants, and several hazardous waste dumps. It is also known for being the naturalist John Muir's home for much of his life. The Martinez Refining Company is the main refinery close to town, operated by Shell. The first major industry to locate in Martinez was a copper smelter and fertilizer plant, built on a rise next to the Bay called Bulls Head Point, a site that later became a sulfuric acid plant and is now owned by Rhodia, a European chemical company. Farther east is the Golden Eagle Refinery, built in 1913 to process heavy crude that comes to the refinery by pipeline from Kern County in southern California. It is operated now by Tesoro. Next to it is a Monsanto chemical plant. Around these sites is a network of chemical landfills, remediation sites, treatment plants, wildlife marshes, and gun clubs.

PORT CHICAGO

Port Chicago is the name associated with the tidal portion of the 13,000-acre Concord Naval Weapons Station, the largest coastal munitions storage facility in the West, though much less active than in the past (and now officially called the Military Ocean Terminal Concord). Part of the site has recently been transferred from the navy to the army, and navy portions are now administered by the Seal Beach Naval Weapons Station in Orange County. The base consists of three sections: the 7,648-acre tidal side, on the water, used for storage and loading, including a separate highly secure area, possibly used for nuclear weapons; the administration/personnel area, on the north side of Highway 4, with barracks, equipment shops, recreational facilities, and offices; and the inland area, on the south side of Highway 4, with 5,272 acres for weapons storage, which has recently been transferred to the army. Port Chicago is best known for an accidental explosion in 1944, at that time the most powerful single explosion to date, in which 320 people loading a munitions ship at Pier 1 were killed. Remains of the pier and a monument are visible on site.

PITTSBURG POWER PLANT

Around the bend from Port Chicago, Suisun Bay's industrial shoreline continues with the largest power plant on the Bay, the 2,000-megawatt natural gas–fired Pittsburg power plant, built in 1954 and expanded in the 1970s with the addition of the 700-megawatt Unit 7. Two other Calpine gas-powered electrical plants in Pittsburg make this the largest energy-producing region in the Bay Area. Much of the industrial port of Pittsburg was transformed during World War II into Camp Stoneman, a 2,800-acre military base that served as a troop training base and part of the San Francisco Port of Embarkation, along with Fort Mason and Angel Island. Stoneman was the largest of these three sites, handling 1.5 million troops during its twelve-year existence. It was closed in 1954, and although it has largely been redeveloped, portions of the base remain. Unexploded munitions have been discovered on Dow Chemical's property and at the fifth hole of the local golf course.

PITTSBURG STEEL PLANT

Pittsburg was named after the steel city in Pennsylvania (although without the *h*) and, like that city, this section of the shore is dominated by heavy industry, including the Bay Area's largest steel plant. The plant opened in 1910 and provided steel for the Bay Bridge and other regional construction landmarks. Its peak employment in the early 1950s, during the Korean War, exceeded five thousand people. In 1986 U.S. Steel joined forces with a Korean steel company to operate the plant as USS-POSCO. Today no new steel is made here; it is a steel shaping, coating, and finishing plant and one of the largest steel facilities on the West Coast.

DOW PITTSBURG

Dow Chemical's Pittsburg plant is the largest integrated chemical plant on the West Coast and a major regional source of herbicides and pesticides. The plant was operating as the Great Western Electrochemical Company when Dow bought it in 1939 in the first major expansion of the company outside its Michigan headquarters area. The site expanded to produce chemicals for military needs in World War II and now has eleven separate manufacturing plants within nearly a square mile of land, each making a different type of chemical product. Products made here are meant to selectively kill microorganisms and macroorganisms, such as Dowicil, an antimicrobial used in cosmetics and paints to inhibit bacterial growth. The plant also makes several pest-control and weed-control chemicals for industrial and consumer products. It employs five hundred people and is one of the company's 197 locations around the world.

PAPER PLANT SITE

Antioch's 3-mile-long shoreline is industrial and postindustrial. At the west end is the old Fulton Shipyard, which closed in 1999. Next to that is a large Georgia-Pacific gypsum plant, which makes wallboard and is the last large manufacturing operation in town. Next to it is the remains of the Gaylord Container plant that made corrugated cardboard for boxes until it shut down in 2002. Opened in 1956 to make paper tissues, towels, and bags from British Columbian pulp, over the years the plant used increasingly more recycled waste paper as a source, eventually becoming one of the largest waste paper processing plants in the nation. The plant has left a legacy of contamination on the shore, including PCBs, and acres of concrete slabs and footings remain behind the locked fence while remediation efforts are addressed.

ANTIOCH POWER PLANTS

Three new gas-fired power plants are clustered at the east end of Antioch, making the 6-mile shoreline between Pittsburg and Antioch the largest source of power for the Bay Area. These three plants were built around the old Contra Costa power plant, an older-style plant from the 1950s, which used river water as its coolant and which has just been retired. The new plants for the most part use evaporative water systems for cooling. Although there are gas fields around Rio Vista in the delta north and east of Antioch, most of the gas for these plants and local industries comes from sources far from the Bay Area, such as Alberta, Canada, and is routed along the southern shore of Suisun Bay after passing through the Antioch Gas Terminal, a mostly underground connection point, located next to the bridge, at the gate of the former DuPont Oakley chemical plant. The Antioch Bridge marks the full transition from the brackish Bay Area to the Sacramento River Delta region.

SUISUN MARSH

Suisun Marsh is one of the largest estuarine marshes in the United States. This 100,000 acres of land and marsh was severely altered by decades of farming, which drained and made levees of much of the area. Farming died out early in the 1900s as the salinity of the available water increased, a problem that also altered the ecology of the marsh. Upstream water diversions to Central Valley agriculture bring about the intrusion of saltwater (when less freshwater comes down the delta, saltwater from the Bay travels farther upstream). Engineering and management efforts are under way to restore freshwater to the marsh. The primary mechanism is the salinity control structure at Montezuma Slough. Built in 1989, its gates are located at the main upstream channel leading into the marsh. They open when freshwater comes downstream from the delta, mostly in the winter, and close when saltwater pushes up from the Bay in the summer.

POTRERO HILLS

The Potrero Hills rise like an island surrounded by the flat expanses of the northern end of Suisun Bay. The hills contain the only remaining explosives plant left in the Bay Area, now owned by UTC Aerospace Systems. The remote hilltop facility, built at a former Nike missile site, employs about 150 people and specializes in small explosive components for aircrafts and missiles. On the same hill, less than a mile away, is the Potrero Hills Landfill, one of the remote regional megafill landfills for the Bay Area. Travis Air Force Base, a major airlift base for the military, lies a few miles north.

MOTHBALL FLEET

The mothball fleet in Suisun Bay is the largest collection of ships on the Pacific Coast of America. These are military ships, the majority of World War II vintage, considered surplus but kept in reserve in case of future need. To help prevent rust and decay, they are stored here, where salt concentrations are less than in the ocean. Two other reserve fleet sites remain in the nation, on the James River in Virginia and at Beaumont, Texas, but these sites are much depleted. The Suisun Bay site has been the largest for some time, counting as many as 342 ships in 1959, but the collection is now rapidly shrinking to meet a 2017 closure deadline set by the government following evidence that the disintegrating ships were depositing tons of contaminants in the Bay from flaking paint. Of the 55 ships on site in 2012, most are likely to be scrapped and towed through the Panama Canal to Brownsville, Texas, currently the only permitted ship breaking yards in the United States, although attempts have been made to approve a facility for demolition at the nearby Mare Island shipyard. Those that are not towed away will be sold, stored elsewhere, or used as targets and sunk at sea.

BENICIA REFINERY

The Benicia Refinery was built by Humble Oil (which became Exxon) from 1966 to 1969 and has the distinction of receiving the first shipload of crude to be delivered from the Trans-Alaska Pipeline, in 1977. Much of the crude processed here still comes from the pipeline, via tanker trips from the pipeline terminus in Valdez, although the refinery also is supplied with oil from the Middle East and is connected to a crude pipeline that brings oil from the San Joaquin Valley. It employs some five hundred people and is considered an average-sized refinery, capable of processing 150,000 barrels of oil per day. When Exxon and Mobil merged, Exxon had to divest itself of some of its assets, including this refinery, which it sold to Valero in 2000. The refinery was built on the grounds of the former Benicia Arsenal, and abandoned munitions igloos can be found on and around it. Unexploded ordnance is still sometimes found and disposed of in the area.

PORT OF BENICIA

What is now the port of Benicia was the industrial center for the arsenal that operated at Benicia from 1849 to 1964, when what remained of its function was transferred to the Tooele Army Depot in Utah. This was the first army arsenal on the West Coast, established to protect the mines and other interests of the newly arriving settlers. Over the years, it served as a manufacturing, service, and storage center for military armaments. During the Korean War it was a repair center for cannons, tanks, and trucks, and it served as a Nike missile maintenance depot for some of the Bay Area's Nike missile sites until the 1960s. When the base closed, ownership of the port areas was transferred to a new company, Benicia Industries, which continues to manage the private port. The largest tenant is the automobile import and logistics company Amports, which can store as many as forty-two thousand cars at Benicia.

MARE ISLAND

Mare Island was once one of the most important navy shipyards in the country, and it is a diversely developed self-contained city. It opened in 1854 as the first American navy base on the Pacific, and during the height of World War II forty-two thousand people worked here. Over its life as a base, hundreds of ships and boats were built on site, including more than a dozen submarines, and thousands of vessels were repaired or refurbished here. It was closed in 1996, and the 5,500-acre peninsula, with nearly a thousand buildings, is being redeveloped. The working waterfront area is the dominant feature of the site, where several dry docks, cranes, and large engineering and assembly buildings continue to be used by civilian companies and military reserve forces. Cleanup of contamination and unexploded bombs in some of the industrial areas continues, with more than twelve thousand ordnance items unearthed so far (not including bullets). The upland magazine area is the first of these sites to be remediated, and it is now a golf course.

SKAGGS ISLAND

For more than fifty years Skaggs Island, at the north end of San Pablo Bay, was a secretive, secure, and self-contained naval facility engaged in a number of communications and intelligence gathering functions. Direction-finding high-frequency antennas at the site helped locate distant sources of communications, as well as the possible location of enemy submarines, by intercepting signals bouncing off the ionosphere. The 3,310-acre site was purchased by the U.S. Navy in 1941 and closed in 1993, but the antennas continued to be used for some time. A staff of a few hundred people was stationed here. The site sat vacant for more than a decade, with the 150 or so buildings, primarily housing, slowly falling apart. Some of the transmitter block houses were used for Navy SEAL training. In 2010 the buildings were demolished in preparation for the site to be transferred to the U.S. Fish and Wildlife Service, which is returning it to marshland.

HAMILTON FIELD

Hamilton Field was among the first major bases to be closed in the Bay Area, in the post–Vietnam War era. Constructed around 1930 as a bomber base, it became a fighter plane and air defense base after World War II and a refugee center during the Vietnam War. The air force ceased many of its functions here by 1974, and by 1976 the base was closed and on caretaker status; debates about its continued use as an airport held up redevelopment plans. A public sale of much of the property occurred in 1985, and the military activities here ceased in 1988. Portions of the base have been given to the city of Novato and developed into housing by private developers. A row of nine hangars remains, most of which have been converted to nonaviation commercial functions, and the runway has been removed and returned to marshland. Two hangars still house a unit of the Coast Guard's national strike force, and the munitions areas of the base are yet to be redeveloped.

POINT SAN PEDRO

Point San Pedro is an old industrial peninsula along the Marin shore. Two local family-owned and -operated bulk material companies, the McNear brickyard and Dutra's San Rafael Rock Quarry, are located next to each other. The Dutra Company is a major mover of material around the Bay. The company started with clamshell dredges working in the delta a hundred years ago and is now involved in marine construction, bridge repair, levee maintenance, dredging, and other shoreline engineering projects. The San Rafael Rock Quarry is the only remaining waterside aggregate quarry on the Bay and a major source of bulk material for local shoreline construction projects. If the quarry closes, it could become a very deep marina. Next to it, McNear brickyard, opened in 1868, is one of the oldest operating brickyards in the state and the last of many that used to be scattered around the Bay.

SAN QUENTIN

This famous state prison is located in Marin County, on the shore of the Bay, at the opposite end of the bridge from Richmond. San Quentin was the first major prison built in California, opening in 1852. It houses the only male death row in California, which, with more than six hundred inmates, is by far the largest condemned population in the nation. The gas chamber, used from 1937 to 1995, is still there, although executions now occur by lethal injection. Like most prisons in the state, San Quentin is overcrowded. Designed for around three thousand inmates, the facility houses more than five thousand.

BELVEDERE ISLAND AND LAGOON

Belvedere is a small residential city located off the Tiburon Peninsula, with an enclosed, man-made private lagoon connecting Belvedere Island to the mainland. With views of the Bay, a large yacht club, and relative isolation from the rest of the world, Belvedere has become, by some ways of measuring, the most affluent place in the nation—it has the highest average individual income for a community with a population over one thousand. A median household annual income of more than $200,000 is no doubt helpful when the average home in the area costs more than $5 million. Similarly hilly and scenic Marin County, which covers the northern shores of San Francisco Bay, including Sausalito, Mill Valley, San Rafael, and Belvedere, is the county with the nation's highest per capita income.

MARINSHIP SITE/SAUSALITO SHORE FRONT

The structure of the shoreline of much of Sausalito today is the result of the Marinship yard, the Bechtel Corporation's massive industrial development from World War II. The company built the 210-acre yard in a few months, filling in tidal areas with soil from an adjacent hillside. Over the next three years, Bechtel employed twenty thousand people to produce ninety-three ships, taking as little as twenty-eight days to make an entire ship. In 1945 Bechtel transferred the site to the Army Corps of Engineers as a base for its postwar Pacific Island reconstruction project. Over the years the site transformed into a commercial and recreational marina area, with a famous houseboat community arising from reuse of the abandoned boats and debris from the old shipyard. Extant remains of the industrial shipyard include shipways, outfitting docks, and a ship mold. The largest building on site houses a large-scale functional model of the San Francisco Bay and Sacramento River system built by the Army Corps in 1957 to study the effects of major engineering projects on the hydrology of the region.

FORT BAKER

The northern side of the Golden Gate is the location of a vista point where the road over the bridge makes its landing overlooking the Bay, and the buildings from the former military base here are known as Fort Baker. Dozens of concrete gun batteries are scattered around the hills, most from World War II (and there are more and bigger ones farther west into the Headlands). The cluster of red-roofed buildings around Horseshoe Bay is the main base, now repurposed as resorts, conference centers, and the Bay Area Discovery Museum for children. The Coast Guard's Golden Gate Station is here, next to underground storage bunkers in the hillside that previously stored floating mines. This station serves the region around the Gate and out past the Farallon Islands. It is very busy, with an average of six hundred search-and-rescue cases every year. Boats from this station are among the first on scene following a report of a suicide off the bridge, an incident that occurs from twenty to forty times per year.

ANGEL ISLAND

Bigger than Alcatraz, Angel Island too is an away place for the city, now as a park but formerly as a site for activities undesirable in the middle of the city. Since its first use as an artillery station in the 1860s, the island has hosted many generations of gun batteries, up until 1962. Meanwhile it served as the Bay Area's primary quarantine and disinfection station, as the immigrant processing and detention center known as the "Ellis Island of the West," and as a prisoner of war camp during World Wars I and II. Now it is operated as a state park, although the ruins of the former uses of the island dominate its atmosphere. A large concrete structure at the East Garrison of former Fort McDowell, known as the Thousand Man Barracks, is one of the earliest forms of tilt-up concrete construction. The top of the island was flattened to house a Nike missile site, but in 2002 the peak was restored by the park service.

ALCATRAZ

Before it became a prison, Alcatraz Island was a fort. At the beginning of the Civil War, in 1861, it had eighty-five cannons and housed more than a hundred men. It soon became a military prison, until 1933, when it was turned over to the U.S. Department of Justice and served as a federal prison until 1963. Ten years later it became a park. Now it is likely the most frequented prison facility in the world, with around 1.3 million visitors annually, who unlike Alcatraz's former population mostly spend less than 2 hours there before returning to the mainland on the ferry.

YERBA BUENA ISLAND

Yerba Buena Island is occupied by the U.S. Coast Guard, which has its San Francisco station here. The base includes the Vessel Tracking Station, with spinning radar and wireless communication towers serving as traffic control for the vessels on and around San Francisco Bay. There is housing for more than a hundred officers (with nice views of San Francisco) and a small port and engineering location for buoy maintenance, security boats, and rescue ships. The Department of Homeland Security opened the Interagency Operations Center, a surveillance and command center focused on activity in the San Francisco Bay region, at the base in 2011. Between 1934 and 1936 the island was drastically transformed when the two spans of the new Bay Bridge were anchored to it, and a 76-foot-diameter, 540-foot-long tunnel was bored through the top of the island for the two decks of the new highway. Construction began at that time as well on Treasure Island, a new landmass created on Yerba Buena's north side.

TREASURE ISLAND

The flat and angular 500-acre, man-made Treasure Island contrasts with the natural and hilly Yerba Buena Island, to which it is connected. Treasure Island was created over two years, starting in 1936, by dumping dredged material on a shoal to house the 1939 Golden Gate International Exhibition, a world's fair that celebrated the joining of the Bay through the recently opened Bay Bridge and Golden Gate Bridge as well as America's growing dominion over the Pacific region. The site was selected because its location in the middle of the Bay meant it did not favor any city, and because it would be a good place for an airport for Pan Am's "flying boats," which were in use at the time. The U.S. Navy took it over soon afterward and built out the island as a training base, which it used for fifty years, until it was officially closed in 1997. Like so many other Bay Area military bases, Treasure Island is slowly being returned to civilian use and now has more than twenty-five hundred residents. Some of the original buildings remain, such as the art deco terminal building and the old China Clipper plane hangars, which have been used for film and television production since the 1980s.

AROUND THE BAY: Man-Made Sites of Interest in the San Francisco Bay Region © 2013 The Center for Land Use Interpretation

Photographs from the Center for Land Use Interpretation photo archive, 2012–2013

The Center for Land Use Interpretation is a nonprofit organization dedicated to the increase and diffusion of knowledge about how the nation's lands are apportioned, utilized, and perceived.

All rights reserved.

No part of this book may be reproduced in any form, by any means, without the express written consent of the publisher.

Text by Matthew Coolidge
Edited by Sarah Simons

The Center for Land Use Interpretation
MAIN OFFICE:
9331 Venice Blvd.
Culver City, CA 90232
www.clui.org

Design and composition by Laura Lindgren

The CLUI would like to thank Laura Lindgren and Ken Swezey, publishers of Blast Books.

Blast Books gratefully acknowledges the generous help of Donald Kennison.

Library of Congress Cataloging-in-Publication Data

Coolidge, Matthew, author.
  Around the bay : man-made sites of interest in the San Francisco Bay region / text by Matthew Coolidge ; edited by Sarah Simons.
     pages cm. — (The Center for Land Use Interpretation American regional landscape series)
  Includes bibliographical references and index.
  ISBN 978-0-922233-43-4 (alk. paper)
1. San Francisco Bay Area (Calif.)—History—Pictorial works. 2. San Francisco Bay Area (Calif.)—Buildings, structures, etc.—Pictorial works. 3. Historic buildings—California—San Francisco Bay Area—Pictorial works. 4. Historic sites—California—San Francisco Bay Area—Pictorial works. 5. San Francisco Bay Area (Calif.)—Description and travel 6. Land use—California—San Francisco Bay Area—History. 7. Industries—California—San Francisco Bay Area—History. I. Simons, Sarah, 1959– editor. II. Center for Land Use Interpretation. III. Title.
  F868.S156C66 2013
  979.4'600222—dc23   2013012319

Published by Blast Books, Inc.
P.O. Box 51, Cooper Station,
New York, NY 10276-0051
www.blastbooks.com

Printed in China

First Edition 2013